Jürgen Schussmann

Design and Cost Performance of WDM PONs for Multi- Wavelength Users

Jürgen Schussmann

Design and Cost Performance of WDM PONs for Multi- Wavelength Users

Future Optical Access Networks

Südwestdeutscher Verlag für Hochschulschriften

Imprint
Any brand names and product names mentioned in this book are subject to trademark, brand or patent protection and are trademarks or registered trademarks of their respective holders. The use of brand names, product names, common names, trade names, product descriptions etc. even without a particular marking in this work is in no way to be construed to mean that such names may be regarded as unrestricted in respect of trademark and brand protection legislation and could thus be used by anyone.

Cover image: www.ingimage.com

Publisher:
Südwestdeutscher Verlag für Hochschulschriften
is a trademark of
Dodo Books Indian Ocean Ltd. and OmniScriptum S.R.L publishing group

120 High Road, East Finchley, London, N2 9ED, United Kingdom
Str. Armeneasca 28/1, office 1, Chisinau MD-2012, Republic of Moldova, Europe
Managing Directors: Ieva Konstantinova, Victoria Ursu
info@omniscriptum.com

Printed at: see last page
ISBN: 978-3-8381-0862-9

Zugl. / Approved by: Ilmenau, TU, Diss., 2008

Copyright © Jürgen Schussmann
Copyright © 2009 Dodo Books Indian Ocean Ltd. and OmniScriptum S.R.L publishing group

Embarking on Health Policy Changes towards Children in Developing Countries

Bharathi Purohit, MDS
Assistant Professor
Department of Public Health Dentistry
Peoples College of Dental Sciences, Bhopal India

Abhinav Singh, MDS
Assistant Professor
Department of Public Health Dentistry
ESIC Dental College & Hospital, New Delhi, India

Acknowledgement

We express our deep sense of gratitude to the **children** who participated in the project. We are grateful to the **school teachers** and the **principals** from various schools for the support and cooperation. We also thank the **Non-teaching staff** for their help and co-operation during the field trips and the survey. In last, we thank **our family** for their constant encouragement and support throughout this project. We would also like to thank God for bestowing his blessings upon us in the form of sweet little angel named Trisha.

Dr. Bharathi Purohit
Dr. Abhinav Singh

Dedicated to our little angel

Contents

1. Introduction	6
2. Aim & Objectives	10
3. Locus of control as oral health correlates	12
4. Knowledge attitude and practices as oral health correlates	29
5. Oral health status among socially disadvantaged children	46
6. Oral health status among special children	57
7. Recommendations	76
8. Conclusion	79
9. References	81

Introduction

Oral diseases qualify as major public health problem owing to their high prevalence and incidence in all regions of the world. According to World Oral Health Report 2003, dental caries and periodontal disease are the two globally leading oral afflictions. The modest goal of reducing DMFT (Decayed, Missing, Filled teeth) to below 3 for 12- year olds and caries prevalence below 50-60% for 5-6 year olds, by World Health Organization (WHO) has not yet been achieved by developing countries. In most of the developed countries including India, these goals are unlikely to be achieved even by the next decade or so, without bringing in a major change in health policies and delivery system.

Barriers exist that prevent many school aged children, not all of whom are poor, from accessing dental care that cannot be overcome by traditional private practice. These barriers include the high cost of fee for service; geographical maldistribution of dentists; disinclination of many dentists to treat poor and minority children. No less significant are the social barriers that include ethnic/cultural attitudes and values, deficient education and inadequate transportation.

In a country like ours, children comprise 40% of a rapidly growing population. Compared to rapid expansions of schools, the provision of health care to these children is poor. Inadequate peripheral health services pose additional problems. Dental health services are poor and rural school dental health services are almost non existent in India. If oral health care of all children have to be taken care, rich and poor alike, then better ways to bring oral health care to children have to be considered. We must acknowledge the obvious fact that with respect to health care, children are essentially non ambulatory. They must have someone with the desire, time money and means to take them to health care provider. If there is no one to bring these children to dental care, then dental care must be provided for them in schools.

School health programs have and are being conducted successfully in developed countries. In developing countries similar programs need to be conducted. The problem being faced is of funding. Who is going to fund these programs? Will it be the government, the school authorities or the public?

Let us first try to understand the problem prior to reaching any conclusions. Developing countries have been consistent in allotting only a fraction of their budget on health. Specifically for oral health, in India there is no existing state or central funding. Ultimate responsibility for the performance of a country's health system lies with the government of that country. We can therefore say that if school health programs have to develop, then it has to be through the government, be it state or central.

The source of funding being clear, it would be up to the policy makers to decide how to bring in the changes. School dental clinics are one of the available options but due to the number of schools, school children and dearth of funds it will not be feasible to appoint dentists in all the schools. Special emphasis need to be given to oral health education and on prevention of oral diseases. Dentist, dental hygienist, dental nurse or even a trained teacher could be involved to conduct educational programs based on the availability of resources. Prior to the policy formulation it is pertinent to be aware of the long outstanding health related problems in the target group. This includes the locus of control, oral health beliefs, oral health behavior and practices and unambiguously the oral health status. Policy formulation requires covering the general population and the more vulnerable groups. Among children the vulnerable group includes the socially disadvantaged and the special care children.

Four different groups of 12 year old children were studied for health locus of control, knowledge attitude practices, and oral health status to determine the existing oral health problems among Indian school children so as to assist in formulation of new health policies. The four groups being:

1. School children in Bhopal city, Central India (Locus of Control)
2. School children from Udupi, Karnataka (Knowledge, attitude & Practices)
3. Socially disadvantaged children in Karnataka, Southern India
4. Special children in Karnataka, Southern India

The government has a duty towards its people, this need to be kept in mind when framing the health policies. How we allocate available funds and services will determine if we succeed in meeting the set goals. Separate funds need to be earmarked for oral health in developing countries. The oral health problems of children cannot be neglected and neither can they be blamed for their poor oral health conditions. If we want school aged children to benefit from adequate oral health care, then it needs to be provided in school health programs, where it can be easily accessed.

Aim & Objectives

Aim of the study:

To craft health policy changes for bringing oral health care towards school aged children

Objectives of the study:

a) *To analyze the health locus of control and to determine its association with oral health among 12 and 15 year old Indian school children*

b) *To determine and characterise the correlation between knowledge, attitudes and practices and oral health among 12 year old Indian school children*

c) *To assess and compare oral health status of 12- year old socially disadvantaged Indian children*

d) *To analyze oral health status of 12 year old Indian children attending special schools*

e) *To provide appropriate recommendations for policy changes and bringing oral health care towards school aged children*

Locus of control as oral health correlates

Health is one of the many areas in which there has been a significant amount of interest in relating locus of control (LOC) beliefs to a variety of relevant behaviors. Locus of control in social psychology refers to the extent to which individuals believe that they can control events that affect them. Health locus of control (HLC) is a construct that refers to how individuals perceive the sources regulating their health. HLC is based on the assumption that health related locus of control scale would provide more sensitive predictions of relationship between internality and heath behaviors. [1]

Locus of control is a product of Rotter's social learning theory. Early HLC studies measured these beliefs on an Internal-External axis. This scale of health beliefs ranged from Internal HLC, where control for one's health resides within the individual, to External HLC, relative powerlessness where control is external to the individual. [2] Individuals with a high internal locus of control believe that events result primarily from their own behavior. Literature indicates that internals are more likely to engage in wide range of health enhancing behaviors than those who believe in chance or social influence on health. Those with a low internal locus of control believe that powerful others, fate, or chance primarily determine events. [3-10]

Better understanding of HLC may enable clinicians to tailor their counseling to suit their patient's health beliefs. HLC is a relatively stable measure in a healthy population.[1] Previous researches have shown that low socioeconomic status, female gender, non-white ethnicity, old age and low education are associated with increased External HLC.[9] Health-related locus of control is significantly associated with a variety of health behaviors and outcomes including knowledge about diseases, ability to stop smoking, adherence to a medical regimen, effective use of birth control, getting preventive inoculations, wearing seat belts, and getting regular dental checkups. [3-10]

There have been few studies reported in the literature on the relationship between locus of control and oral health, but the findings have been contradictory. [9-10] Therefore if the association is understood, then parents, school officials and health professionals can be made more aware of the circumstances that might lead to the adoption of particular locus of control belief which may aid in improving the overall oral health of the community.

As an independent variable, it is important to note that there is no theoretical reason to expect locus of control to predict health behavior. Hence, the purpose of this study was to investigate the correlates of the HLC subscales with oral health status. The specific aim of the study was to analyze the health locus of control and to determine its association with oral health among 12 and 15 year old school children from Bhopal City, Central India.

Methods

Study design and subjects

The target population for the cross-sectional study was 12 and 15 year old school children from Bhopal City, an urban area in Central India. According to the estimated prevalence in the pilot study and assuming a standard error of 5%, the sample size derived was 246 and 196 subjects for 12 and 15 year children. It was decided to select five schools from the sixteen government schools to achieve the desired sample size. Five schools were randomly selected by lottery method and all the 482 children from the study age group were invited to participate in the study. At the end of the study, a total of 462 children were examined; 259 from the 12- year age group and 203 from 15 year age group. The school teacher along with the survey team requested the children to participate, giving a high response rate of 96%.

Questionnaire

The 11-item Health Locus of Control (HLC) Scale developed in 1976 by Wallston, was a health-specific version of Rotter's 1966 Internal-External Scale (I-E scale), which was used to classify individuals as internals or externals.[7] This was actually based on the belief that locus of control is a unidimensional construct. However, later it became clear that locus of control was multidimensional; internality and externality were basically uncorrelated rather than being opposite ends of the same pole. The same individual always scores on both scales (more or less) and that a high score in one scale does not necessarily yield a low on the other. The multidimensional HLC construct is an improvement over the classic unidimensional conceptualization; it measures health beliefs with a tripartite approach by differentiating External HLC into Powerful others

HLC (eg, physicians) and Chance HLC.[8] The three dimensions are traditionally treated as independent factors, though studies have revealed modest between-factor correlations.[4-5] The acceptable validity and reliability of the HLC scale have been well-documented over its 30-year history.[6]

The eleven-item health locus of control questionnaire (HLC) was used to assess health locus of control.[7] There were 5 questions which were used to assess the IHLC (Internal health locus of control) (Questions 1,2,8,10,11) and 6 questions for EHLC (External health locus of control) evaluation (Questions 3,4,5,6,7,9). External HLC comprises of Chance HLC and Powerful others HLC. Three questions was used to assess Chance health locus of control (CHLC) (Questions 3,5,9) and three questions were used for evaluation of Powerful others HLC (PHLC) (Questions 4,6,9). The HLC Scale was scored so that high scores indicated agreement with internally worded beliefs. Individuals with scores above the median were labeled "health-internals"; who believe that the locus of control for health is internal and that one stays or becomes healthy or sick as a result of his or her own behavior. At the other end of the dimension, scoring below the median, were the health-externals, they were presumed to have generalized expectancies that the factors that determine their health are ones over which they have little control; i.e., external factors such as luck, fate, chance, or powerful others.

Based on the pilot study a six point Likert scale was used instead. Respondents rate each item on the HLC using a six point (1 to 6) scale; thus, scale of eleven questions had a scoring range from 6 to 66. Higher subscale indices reflect stronger perception of control in the given dimension. Individuals were classified as internal or external based on the median score; internals scoring above 33 and externals scoring below or equal to 33. Reverse scoring was done for questions assessing external locus of control scores.

Till date there are no reports on the validation of an Indian version of HLOC questionnaire. English version of the questionnaire was used in the study as all the schools in the city are English medium and / or have English as a compulsory subject. The questionnaire was simple enough to be understood by the children of the selected age group. The questionnaire was used and pretested on a random sample of school children to ensure practicability, validity and interpretation of responses. The reliability of questionnaire was assessed using Cronbach's alpha internal consistency coefficient.

Clinical examination

All the subjects were examined under adequate illumination (lighting conditions). Type III examinations were used for data collection. Clinical examination using a mouth mirror and dental probe under adequate illumination is referred to as Type III examination.[11] Clinical data was collected for oral hygiene and dental caries status.

Simplified oral hygiene index (OHI-S) developed in 1964 by John C. Greene and Jack R. Vermillion was used to evaluate oral hygiene status.[12] OHI-S has two components, the simplified debris index (DI-S) and the simplified calculus index (CI-S). Each of these indices, in turn is based on numerical determinations representing the amount of debris or calculus found on the preselected tooth surfaces. The six surfaces examined for the OHI-S are selected from four posterior and two anterior teeth. The designated tooth to be examined were 16,11,26,36,31,46. Only fully erupted permanent teeth were scored. Natural teeth with full crown restorations and surfaces reduced in height by caries or trauma were not scored instead an alternative tooth was examined.

Alternative teeth for first molars were second and third molar of the same quadrant; and adjacent laterals for the central incisors. For each individual, the debris and calculus scores were added and divided by the number of tooth surfaces recorded. The OHI-S was interpreted as: good (0 to 1.2), fair (1.3 to 3) and poor (3.1 to 6) ranging from 0-6.

WHO's criterion was used for detection of dentition status. [13] The examination was conducted with a plane mouth mirror and a dental probe. A systematic approach was adopted for assessment of dentition status. The examination proceeded in an orderly manner from one tooth or tooth space to the adjacent tooth or tooth space. A tooth was considered present in mouth when any part of it was visible. If permanent and primary tooth occupied the same tooth space, status of permanent tooth was recorded. Decayed, missing and filled teeth scores were calculated using DMFT index. The D component was used to describe decayed teeth. Caries was recorded as being present when a lesion in a pit or fissure or on a smooth surface had a detectable softened floor, undermined enamel, softened wall or a temporary filling. On proximal surfaces, the probe had to enter a lesion with certainty. A tooth was considered missing, if there was a history of extraction due to pain and / or the presence of a cavity. The M component was used to describe missing teeth due to caries. The F component was used to describe teeth that have been previously filled as a result of caries involvement.

The survey was scheduled between the months of October 2010 and Feb 2011. Information on demographic characteristics of participants were collected by means of personal interviews administered by the examiner. All examinations were performed by a single examiner and duplicate examinations were conducted on one of every ten subjects throughout the survey. The dental team comprised of the examiner assisted by a recording clerk.

Training and calibration

Three day training session for standardization and calibration of data collection methods were organized. Training session consisted of review of the criteria used followed by clinical examination. Calibration of examiner was conducted by examining a total of 20 children twice for oral hygiene and dental caries, with a 60 minutes interval between the examinations. Intra- examiner reliability was assessed using kappa statistic which was in range of 0.88 – 0.92 for both the parameters studied, showing a high degree of conformity in the observations. Recorder was instructed in the coding systems of the indices used in the study. The recorder subsequently practiced these by recording findings during the calibration of examiner.

Informed consent

University clearance was granted for the study. Permission was sought from district educational officers and from Principals of selected schools. Informed written consents were taken from parents and school children prior to carrying out the survey.

Statistical analysis

All collected data were entered into spreadsheets. SPSS software version 16 was used for statistical analysis. Chi Square test was used to compare between categorical variables. Internal and external health locus of control scores were calculated. Mann-Whitney test was used for comparisons of health locus of control scores (Internal and External) with quantitative variables. Mean OHI(S) and DMFT scores were calculated for the two target age groups. Kendall's tau-b correlation tests were used to correlate health locus of control scores (Internal and External) with OHI-S and DMFT scores. $p < 0.05$ was considered as statistically significant.

Results

A total of 462 children were examined; 259 were from the 12- year age group and 203 children from 15 year age group. The number of boys and girls were 273 (59%) and 189 (41%) respectively (Table 1). Table 2 presents mean health locus of control scores for each question.

A minimum mean score of 2.89 ±1.58 and 1.89 ±1.27 and a maximum mean score of 4.65 ±.95 and 4.70 ± .87 were recorded for the 12 and 15 year age groups respectively. Mean OHI (S) score of 1.56 ± 1.94 and 1.60 ± 1.67 for the 12 and 15 year age groups were recorded. Mean DMFT scores of 1.26 ± 1.32 and 1.34 ± 1.44 were also revealed for the two age groups respectively. A non significant increase in IHLC and EHLC scores, OHI (S) and DMFT scores were noted when comparing 12 and 15 year age groups (Table 3).

A negative significant correlation was noted between internal health locus of control and DMFT scores ($r = -0.42$; $p < 0.001$). Significant positive correlation was noted between EHLC and OHI (S) scores ($r= 0.15$; $p <0.05$) and DMFT scores ($r = 0.38$; $p < 0.001$) (Table 4). The results demonstrate that as the score increases in the external belief scale there is also a slight increase in OHI (S) scores. The level of this correlation is weak therefore a low r value and a week statistical significance.

Table 1: **Age and Gender distribution of study subjects**

Age group	Gender		Collective
	Boys (%)	Girls (%)	
12 years	151 (58.3%)	108 (41.7%)	259
15 years	122 (60%)	81 (40%)	203
Total	273 (59%)	189 (41%)	462

Table 2: **Health Locus of Control among study subjects**

HLC Questionnaire	Mean (SD) (Based on 6 point scale)*		
	12 years	15 years	Collective
1. If I take care of myself, I can avoid illness.	4.65 (0.95)	4.70 (0.87)	4.66 (0.96)
2. Whenever I get sick it is because of something I've done or not done.	3.32 (1.10)	3.58 (0.93)	3.43 (1.03)
3. Good health is largely a matter of good fortune.	4.57 (1.07)	4.02 (1.41)	4.33 (1.26)
4. No matter what I do, if I am going to get sick I will.	3.15 (1.26)	2.94 (1.38)	2.94 (1.38)
5. Most people do not realize the extent to which their illnesses are controlled by accidental happenings.	3.15 (1.28)	3.34 (1.31)	3.34 (1.31)
6. I can only do what my doctor tells me to do.	4.22 (1.15)	4.21 (1.0)	4.21 (1.08)
7. There are so many strange diseases around that you can never know how or when you might pick one up.	3.15 (1.26)	3.94 (1.19)	3.50 (1.26)
8. When I feel ill, I know it's because I have not been getting the proper exercise or eating right.	4.16 (1.12)	4.01 (1.22)	4.09 (1.17)
9. Whether you keep your teeth or lose them is mostly a matter of luck.	2.89 (1.58)	1.89 (1.27)	2.45 (1.53)
10. Bad oral health results from one's own carelessness.	4.22 (1.32)	4.46 (0.97)	4.32 (1.18)
11. I am directly responsible for my oral health.	4.35 (1.06)	4.53 (0.94)	4.52 (0.94)

* Minimum possible mean score: 1, Maximum possible mean score: 6

Table 3: Mean Health locus of control (HLC), OHI(S) and DMFT among study subjects

Variables	12 years	15 years	Collective	p Value
Internal HLC	4.15 ± 1.48	4.24 ± 1.21	4.19 ± 1.34	0.81
External HLC	3.39 ± 1.78	3.48 ± 1.21	3.43 ± 1.49	0.52
i) Powerful HLC	4.20 ± 1.48	4.21 ± 1.02	4.21 ± 1.25	0.74
ii) Chance HLC	2.58 ± 2.08	2.75 ± 1.4	2.67 ± 1.74	0.48
Distribution of study subjects based on Health Locus of Control				
Variables	12 years	15 years	Collective	p Value
Internal HLC n (%)	199 (76.8%)	147 (72.4%)	346	< 0.001**
External HLC n (%)	60 (23.2%)	56 (27.6%)	116	< 0.001***
Mean OHI (S) and DMFT scores among study subjects				
Variables	12 years	15 years	Collective	p Value
Mean OHI (S)	1.56 ± 1.94	1.60 ± 1.67	1.58 ± 1.91	0.65
Mean DMFT	1.26 ± 1.32	1.34 ± 1.44	1.30 ± 1.38	0.48

* $p < 0.05$ (statistically significant but weak associations)

** $p < 0.01$ (highly statistically significant)

*** $p < 0.001$ (very highly statistically significant)

Table 4: Correlation between Health locus of control (HLC), OHI (S) and DMFT

Variables	Correlation coefficient (r)	p Value
Internal HLC-DMFT	- 0.42	< 0.001***
External HLC -DMFT	0.38	< 0.001***
Internal HLC-OHI (S)	- 0.07	.086
External HLC – OHI (S)	0.15	< 0.05*

*p < 0.05 (statistically significant but weak associations)

**p < 0.01 (highly statistically significant)

***p < 0.001 (very highly statistically significant)

Discussion

In the present study the health locus of control was assessed among 462 children aged 12 and 15 years in Bhopal city, using eleven-item health locus of control questionnaire (HLC). The results reveal a high Internal locus scores among both the age groups; higher scores being noted among 15 year olds when comparing the two age groups.

Acharya (2008) conducted a study to appraise the effect of different stages of professionalization on the health locus of control among Indian dental students and concluded that the mean score for the Internal subscale was consistently higher than the mean scores for external HLC. [14] Literature suggests that the health locus of control beliefs can be used to predict health practices and outcomes in the long run, that they are amenable to change, and that those who report more internal health locus of control are more likely to proactively seek health-promoting information and skills, realize for themselves the link between their lifestyle and health, and purposefully engage in initiatives associated with psycho-social and developmental well-being. [15] In the present study a non-significant increase in OHI (S) and DMFT scores were noted when comparing the two age groups. Significant positive correlation was noted between external health locus of control, OHI (S) scores and DMFT scores. Chase et al (2004) [3] and Brandao et al (2006) [10] reported no relationship between oral health status and locus of control while significant associations similar to the present study were reported by Acharya et al (2011) [16] and Lencova et al (2008). [17] Self-rated oral health, socio-demographic factors, and oral health behaviors were significantly associated with oral health control beliefs in a study conducted by Peter and Bermek (2010) using multi dimensional oral health locus of control scale. [18]

Researchers have linked internal locus of control to positive health beliefs and behaviors, yet not all attempts to correlate the two have been successful. [16] Therefore, rather than associating the already proven oral health behaviours and oral health, the present study was designed to assess, whether or not, the general health-related locus of control is also associated with oral health outcomes. That is the reason also why we chose to measure controllability in a general way (i.e. Health Locus of Control) rather than in a more specific way (i.e. perception of control over dental behaviors). Since oral health is a part of general health and targeting general health behavior may also lead to overall health improvement of an individual; including oral health. This may seem even more appropriate for countries with deficient healthcare workers.

Wolfe et al (1996) reported a shift in locus of control, from external to internal as a result of oral hygiene interventions. [19] Therefore providing children with advice and reinforcements may change their perceptions of controllability. [20] English version of the questionnaire was used in the study as all the schools in the city have English as a compulsory subject and the students had a good knowledge in the language. Confounders such as oral hygiene practices, sugar consumption and utilization of dental care have not been considered separately as they directly or indirectly are associated with health belief and behavior of an individual. [10,14,17] In an effort to further investigate interactions of the risk factors involved in the etiology of oral diseases, research has focused on socioeconomic factors as these could act as indirect causal agents. When considering socio-economic indicators as risk factors, it has been recognized that children's oral health is related to their families' socioeconomic status (SES) and their mother's education level. [19] The study was conducted in mixed areas of the city. Although, government schools mainly attract children from lower and middle SES, yet the possibility of few belonging to higher SES cannot be ruled out. The effects of these factors were not taken in consideration and could be considered as the limitation for the study. Results are compared with studies which consist of populations of other ages and disease groups therefore it is also important to consider these differences.

The health locus of control scale is recommended in conjunction with behavioral measures to evaluate the success of health education programs. Since it is true that internal's appear more likely to engage in positive health and sick-role behaviors, it is apparent that the health locus of control emphasizes the importance of the health educators need to involve themselves in training patients to hold more internal beliefs.

A plethora of evidence linking the growing recognition that the school system has an important role to play together with the family and community in enhancing favourable health and developmental outcomes for young people, suggest that education content, methods and values can provide necessary traction and leverage for promoting adolescents' self-development and thriving in the world. [21] It is logical to expect that school teachers could help shape perceptions of internal (health) locus of control and enhance protective resources among school-aged adolescents by building on students' strengths and resources with timely and supportive feedback; to help students maintain good oral health. Internal's appear more likely to engage in positive health and had better oral hygiene and lower dental caries levels. These beliefs may be useful for planning oral health promotion programs and for formulating advice given by oral health professionals about their patients' oral health behaviors.

Knowledge attitude & practices as oral health correlates

Over the past few decades a marked decline in dental caries experience of children has been observed in many industrialised countries. In developing countries the trend indicates an increase in oral health problems; specifically dental caries.[1] In parallel with the changing oral disease patterns, there has been significant improvement in the oral health awareness, dental knowledge and attitude of the children and parents. Conversely, increasing levels of dental caries have been observed in several developing countries, especially for those countries where preventive programs have not been implemented.

The oral health problems that are currently most prevalent - dental caries, periodontal diseases, and dental trauma - can be prevented by measures aimed at reducing the exposure to risk factors. However, such prevention requires subjects to be adequately informed about the causal factors. Studies suggest that a high proportion of the population have limited understanding of how to prevent oral diseases.[2]

This aspect is a starting point for health promotion campaigns. The affected population needs to receive information on oral diseases, risk factors and measures that can be adopted to prevent them. Such campaigns will typically aim not only to impart knowledge, but also to improve the attitudes regarding oral health, and to facilitate the transformation of these attitudes into practice.[3]

Within the KAP (Knowledge, Attitude and Practice) model, the change from an unhealthy attitude to a healthy attitude will occur when given adequate information, adequate motivation and adequate practice of the measures to be adopted by the subject.[3] Information means that the subject has all the data necessary to understand what oral disease is and how it arises, as well as to understand the protective measures that need to be adopted. This knowledge will, in theory, lead to the changes in attitude, which will in turn lead the subject to make changes in their daily life. Thus in the case of dental caries, the subjects know that incorrect brushing may cause caries, and this information generates a positive attitude towards daily brushing (i.e. the intention to brush teeth daily in order to have healthier teeth), and thus the changes in brushing behaviour.

The KAP model of behavioural changes is in fact solidly embedded within the traditional focus of health education. It is a model with a positive vision of science, treating the behavioural change as a logical individual decision: the individual can be expected to change an unhealthy habit to a healthy habit in the light of information on the health benefits of that change. This theory considers that individual factors are the principal determinants of disease, biological or behavioural.[4-6] However, the model must be considered incomplete in terms of the practical application in health education, since it does not take into account the subject's environment and sociocultural context.

The importance of knowledge, attitude and practices cannot be overemphasized. Various studies have been conducted to appraise the oral health knowledge, attitude and practices but none have correlated them as suggested in KAP model. Keeping this in view, the present study was conducted with objectives of assessing knowledge, attitude and practices of oral health among 12- year old schoolchildren in Udupi, South India and to correlate this with the oral hygiene and dental caries status of the children.

Methods

Study design and subjects

The target population for the cross-sectional study was 12- year old school children of Udupi district, South India. List of the schools along with the numbers enrolled was obtained from the Board Development Office, Udupi. Permission to conduct the study was obtained both from the school authority or principal.

Prevalence of dental caries among school children was taken to be 50% based on pilot study results. Using the formula, n= Z^2 PQ/ d^2, at 95% confidence Interval and standard error (d) of 5, sample size was calculated as 384. Hence eight government and private schools were randomly selected and all the children in these age groups were examined following the inclusion and exclusion criteria, giving a final sample size of 412.

Children in the age of 12 ± 1 years and those willing to participate were included in the study. Children absent on the day of examination, those undergoing orthodontic treatment and children with medical condition that contraindicates oral examination without appropriate modifications e.g. infective endocarditis were excluded from the study.

Children in the age range of 12 ± 1 years studying in 6th class were considered for the study (as recommended in Manual for Multi-Centric Oral Health Survey compiled by Centre for Dental Education and Research, All India Institute of Medical Sciences, New Delhi and World Health Organization, 2004 - 2005).[7]

A questionnaire was prepared to assess the knowledge, attitude and practices of children with the help of experts in the field. The questionnaire was pretested and questions framed were keeping the study group in mind. Validation of the questionnaire was assessed using the Cronbach's alpha. The questionnaire consisted of three parts. There were 7 questions which assessed the knowledge, 3 questions to assess the attitude, and last 6 questions assessed the oral health practices of the schoolchildren. School children were distributed the questionnaire in their respective classrooms and asked to fill, without discussing with each other, in 15 minutes. The subjects were advised to provide correct information about the asked questions and not to over or under report.

Clinical Examination

A schedule was prepared for the data collection. 20 – 25 school children were to be examined per day. The survey was scheduled during the months of January and February 2007. A single examiner, trained and calibrated for the indices used conducted the clinical examination in the school premises. Examination was carried out in the natural day light conditions, with the subjects seated on a chair with backrest and the examiner standing behind the subjects. Mouth mirror and an explorer were used in examination procedure. OHI-S (1964)[8] was used to assess oral hygiene status since it represents both the debris and calculus (proxy indicators of oral hygiene) in one score thereby permitting correlation of knowledge, attitude and practices scores separately with oral hygiene. DMFT (1986)[9] was used to assess caries status.

Ethical clearance was taken from the Kasturba Hospital Ethics Committee, Kasturba Hospital, Manipal. Informed written consent was taken from parents and children before carrying out the survey. All examinations were performed by a single examiner and duplicate examinations were conducted on one of every ten subjects throughout the survey. Intra- examiner reliability was assessed using kappa statistic which was in range of 0.82 – 0.84.

Statistical analysis

Each correct answer was given a score of 1. Incorrect answers or those not following the recommended guidelines were not given any score. Knowledge, attitude and practice scores were calculated separately for each subject. These scores were correlated individually and with the OHI-S and DMFT scores. Kendall's tau b test was used to compute correlation between the various scores. Mann – Whitney U-test was used for comparison between two groups for the quantitative variables. The significance value was fixed at a p value of < 0.05. Statistical analysis was performed using the Statistical Package for Social Sciences (SPSS) software for windows version 16.

Results

Table 1 shows distribution of schoolchildren based on sex, type of school and medium of instruction. The number of subjects selected was 412; 280 were boys and 132 were girls. There were 87 children from the government school and 325 from the private schools, out of which 115 had Kannada and 297 had English as their medium of instruction.

Majority of the schoolchildren (54.8%) knew that tooth brushing helps in preventing caries. Around forty two percent of the subjects were aware that tooth brushing helps in preventing periodontal diseases. Only 38.5 % of the students knew what to do after dental injury. Majority (95.2%) students reported fruits like apple cause decay of teeth and 89.4% children thought fruit juices to be most harmful to teeth.

Almost all the children (98.1%) were interested to know more about how to keep the teeth clean. Majority of the candidates (88.5%) wanted to know more about what to do in case of dental injury. Thirty one percent children felt that they themselves are responsible for their dental treatment whereas 61.5% felt parents are responsible for it. (Table 2)

Majority (91.3%) reported that they use toothpaste and toothbrush for cleaning teeth. Substantial number of children (43.3%) did not know whether they use fluoridated toothpaste or not. 60.6% children never visited a dentist in past one year, and among those who visited 58.5% went for general check-up and 22% visited with the complaint of pain. 49% children admitted to eating sweets very often.

Mean OHIS score of the study population was 1.16 ± 0.837 (Table 3). Mean DMFT score of the study population was 1.86 ± 0.62. The findings were non significant when compared to various groups. (Table 4)

Table 5 depicts the correlation between knowledge, attitude, practice, OHI-S and DMFT scores. Significant correlation was seen between Practice- OHI-S and Practice- DMFT scores.

Table 1: Distribution of subjects according to gender, type of school and medium of instruction

Variables		Number	Percentage %
Gender	Male	280	68
	Female	132	32
School	Government	87	21.2
	Private	325	78.8
Medium of instruction	Kannada	115	28
	English	297	72

Table 2: Frequency table for knowledge attitude and practice

Question	Response	Number	Percentage %
Tooth brushing helps in preventing caries?	Yes	225	54.8
	No	15	3.8
	Don't know	172	41.7
Tooth brush helps in preventing gum diseases?	Yes	174	42.3
	No	99	24
	Don't know	139	33.7
Do you use fluoridated tooth paste?	Yes	186	45.2
	No	47	11.5
	Don't know	179	43.3
Do you know what to do after dental injury?	Yes	159	38.5
	No	253	61.5
Have you been to dentist in past 1 year?	Yes	162	39.4
	No	250	60.6
Who is responsible for your dental appointments?	Yourself	131	31.7
	Parents	253	61.5
	Teachers	20	4.8
	Dentists	8	1.9

Table 3 a: Mean OHI-S scores in study population

Variables	Mean OHI-S score	p value*
Gender	Male 1.19 ± 0.845	0.652 not significant
	Female 1.15 ± 0.651	
School	Government 1.24 ± 0.672	0.421 not significant
	Private 1.15 ± 0.864	
Medium of instruction	Kannada 1.16 ± 0.977	0.976 not significant
	English 1.16 ± 0.781	

*Mann – Whitney U-test

Table 3 b: Oral hygiene scores in study population

Oral hygiene scores	Number of subjects (%)
Good	238 (57.7%)
Fair	84 (20.3%)
Poor	90 (22%)

Table 4: Mean DMFT scores in study population

Variables	Mean DMFT score	p value*
Gender	Male 1.96 ± 0.84	0.53 not significant
	Female 1.80 ± 0.65	
School	Government 1.86 ± 0.37	0.40 not significant
	Private 1.76 ± 0.86	
Medium of instruction	Kannada 1.80 ± 0.57	0.92 not significant
	English 1.75 ± 0.78	

*Mann – Whitney U-test

Table 5: Correlation between various scores in study population

Variables	Correlation coefficient	p value*
Knowledge - Attitude	0.140	0.21
Knowledge - Practice	0.082	0.181
Attitude - Practice	0.182	0.19
Knowledge – OHI-S	0.030	0.182
Attitude – OHI-S	0.052	0.885
Practice – OHI-S	-0.889	**0.01**
Knowledge - DMFT	0.130	0.188
Attitude - DMFT	0.002	0.985
Practice - DMFT	-0.765	**0.01**

*Kendall's tau b test

Discussion

The present survey was completed to congregate epidemiological data for associating knowledge, attitude and practices of oral health and also to ascertain its association with the oral hygiene and dental caries status.

Various studies[3,10,12,13,14,16] have been conducted to appraise the knowledge, attitude and practices but none in literature have correlated the scores with oral hygiene and DMFT scores. In the present study, majority of the subjects, were not aware whether they used fluoridated toothpaste, similar to the findings of Varenne et al [10] who conducted a study to assess oral health behaviour of children in 12 years old in urban and rural areas of Burkina Faso, Africa. Comparable findings were also reported by Kassim et al [11] among 141 Kenyan adult population in a rural arid setting.

Regarding dental visits, about 60% of subjects had never seen a dentist in past 1 year which is quite high as compared to a study by Zhu et al [12] in children and adults of age groups 12 and 18 years, where about one third of participants had never seen a dentist. The most frequent reason for visit was general check-up, and pain. In another study by Penq et al [13] among 12-year old urban schoolchildren in the People's Republic of China only 46% schoolchildren had seen a dentist within the past year.

Majority of the candidates did not know whether dental floss helps in preventing caries, periodontal disease and tooth brushing prevents periodontal disease. Most of the subjects knew (54.8%) tooth brushing prevents caries. Lack of knowledge concerning dental floss and periodontal diseases can be attributed to the low level of dental health education of the school children. 89.4% of the candidates considered fruit juices more harmful to the teeth than soft drinks and fruits like apples to cause the decay of teeth more than the bread and chocolates.

Petersen et al [14], in a study conducted among 12 year old urban and rural schoolchildren in Southern Thailand concluded that the consumption of sweets is an important predictor of the high caries experience. In the present study 49% children reported of sweet consumption very often and 14.4% reported of its consumption all the time.

The findings indicate that most of the children lacked correct knowledge about the causes and prevention of dental diseases, as seen in other similar studies. Al-Hussaini et al 2003[15] evaluated the dental health knowledge, attitudes and behaviour among students at the Kuwait University Health Sciences Centre (HSC). His findings indicate that although most of the students at the HSC seem to be satisfied with their dental health, they did not have the correct knowledge about the causes and prevention of dental diseases.

The result of the present study showed that the subjects appeared to be interested to receive more information on a substantial number of different dental issues. Significant number of candidates wanted further information, so it can be said children had a positive attitude towards oral health.

Christensen et al 2003 conducted a study to assess the prevalence of dental caries, and to assess the level of attitude, knowledge and practice in relation to oral health and oral health behaviour among 11-13 year-old children in Bhopal, India.[16] They concluded that the implementation of community-oriented oral health promotion programmes is needed in order to increase the level of knowledge and to change attitudes and practices in relation to oral health among children.

There was no correlation seen between knowledge, attitude and practice scores as suggested in KAP model. Significant correlation was noted between the Practice- OHIS scores and Practice-DMFT scores. This finding points to the significance of correct practices to maintain good oral health.

This project was dependent on self-reported data derived from the school children with varying levels of familiarity with completion of questionnaires and varying levels of language ability, which may have influenced the selection of the responses. This might limit the study due to misinterpretation and misunderstanding of questionnaire items by the subjects. However, the questionnaire was pretested before the study was conducted with positive results. Furthermore, the investigator was always available during the completion of the questionnaire, and the subjects were encouraged to approach him whenever they needed clarification of any point. With regard to attitudes towards the teeth and dental care, oral hygiene habits and frequency of dental visits, over reporting has to be assumed whereas the consumption of sugary foods and drinks has probably been under reported. In addition, recall bias should be considered with respect to consumption of food items and services received at the last dental visit. The questions on oral health related knowledge, attitude and practices employed were assumed to be sufficiently simple to achieve a reasonable degree of validity and reliability.

Knowledge about the periodontal diseases and use of fluoride was found to be low, dental visits were infrequent, and sweet consumption was found to be high. There was no correlation between the knowledge, attitude and practice scores. Significant correlation was seen between the Practice- OHIS and Practice-DMFT scores.

Oral health status among socially disadvantaged school children

Oral diseases as stated earlier have emerged as major public health problem owing to their high prevalence and incidence around the world. According to World Oral Health Report 2003, dental caries and periodontal disease are the two globally leading oral afflictions.[1]

From a number of point prevalence studies, and National Oral Health Survey and Fluoride Mapping 2002-2003 a fact emerges that dental caries and periodontal disease has been increasing both in prevalence and severity over the last decades.[2-8]

Early recognition of the disease is of vital importance, in order to prevent the disease and pain so as to make oral health services more relevant. Early intervention for better health depends on early identification. Timely detection is essential to alert both parents and health care professionals to for appropriate action as the disease is expensive to treat and many patients cannot afford to take or complete treatment for dental caries as recommended by their dentist.

In India children comprise 40% of a rapidly growing population. Compared to rapid expansions of schools, provision of health care to these children is poor. Inadequate peripheral health services pose additional problems. Dental health services are poor and school dental health services are almost non existent in India.

Society and culture are linked to behavioral patterns or lifestyles. Therefore, there is a need to explore the influence of social factors on oral health. One of the measures of social differentiation is the area of residence i.e. rural and urban people. Another important way of distinguishing people is by their socioeconomic status. It is suggested that low socioeconomic status individuals have fatalistic or pessimistic views of health in general and oral health specifically. [9] Various indicators of socioeconomic status such as income, education, occupation have been used.

Amongst school children, different socioeconomic groups have been identified based on different school types attended.

Ashrama schools are government schools for children of backward classes i.e. Scheduled Caste (SC), Scheduled Tribe (ST), and Other backward classes (OBC). These residential schools charge no fees; instead parents are given financial incentives to educate their children, thus attracting children from lower socioeconomic strata.

This study explores the association between social disadvantage and oral health. Specifically, the aim of the study was to compare and assess oral health status of 12- year old children from two socially disadvantaged groups in Udupi District, South India.

Methods

Study design and subjects

The cross-sectional study was conducted among 12- year old children residing in Ashrama schools in Udupi district - Karnataka State, India. List of Ashrama schools was obtained from the Indian Tribal Development Office, Udupi. All six Ashrama schools in the district were covered. A total of 327 children were attending these schools; all children were examined.

For the purpose of comparison with children of Ashrama schools, 4 non- residential unaided government schools were randomly selected. A total of 340 children in age group 12- years were selected for comparison from other government schools. Children in age range of 12 ± 1 studying in 7th class were considered for study (as recommended in Manual for Multi-Centric Oral Health Survey compiled by Centre for Dental Education and Research, All India Institute of Medical Sciences, New Delhi and World Health Organization, 2004).

A proforma was designed to record information regarding demographic data, oral hygiene practices, dietary habits and last dental visit. Single examiner trained and calibrated for criteria conducted the clinical examination. Dental fluorosis, periodontal status as per Community Periodontal Index (CPI), dentition status (DMFT) and dentofacial anomalies using Dental Aesthetic Index (DAI) were included in modified WHO Oral Health Assessment form (1997). [10]

Survey was scheduled during the months of April and September 2007. Pilot study was conducted among 30 children in comparable age groups. Intra- examiner reliability was assessed using kappa statistic which was in range of 0.78 – 0.82.

Ethical clearance was taken from the Kasturba Hospital Ethics Committee, Manipal. Children requiring immediate care were referred to Manipal College of Dental Sciences, Manipal or to the nearest dental hospital.

Data was transferred from pre - coded proforma to computer and analysed using SPSS version 13. Chi Square test was used to compare between categorical variables. Mann – Whitney U-test was used for comparison between two groups for quantitative variables. $p \leq 0.05$ was considered as statistically significant.

Results

A total of 667 children comprised the sample. Among 327 children examined in Ashrama schools, 157 (48%) were males and 170 (52%) females. In comparison group, of the 340 children examined 146 (42.8%) were males and 194 (57.2%) females. Significant differences were found between two groups with respect to frequency of cleaning teeth, material used for cleaning, type of diet, sugar consumption and dental visits. (Table1)

Dental fluorosis was present in 75 (22.9%) children from Ashrama schools, whereas in the comparison group 49 (14.4%) children had dental fluorosis ($p \leq 0.001$). (Table2) Out of 74 (100%) children with dental fluorosis in the Ashrama school, 10 (13.7%), 31 (41.3%), 29 (37.9%) and 5 (6.9%) children had questionable, mild, moderate and severe fluorosis respectively. Out of 49 (100%) children in the comparison group, 5 (10.5%), 31 (63.3%), and 13 (26.3%) were having questionable, mild and moderate fluorosis .

Calculus was present in 260 (79.5%) children from Ashrama schools as compared to 187 (55%) children from the comparison group ($p \leq 0.001$). (Table2)

Mean number of Decayed teeth and mean DMFT values in children of Ashrama schools were 1.15 ± 1.62 and 1.15 ± 1.62 respectively. In comparison group the values were 0.46 ± 0.98 and 0.48 ± 1.04. No significant differences were noted between two groups with respect to Dental Aesthetic Index (DAI) scores. (Table2)

Table 1: Demographic and oral health behavioral characteristics of study population

Oral health related behavior variables		Ashrama School N (%)	Govt. School N (%)	p value
Gender	Male	157 (48)	146 (42.8)	
	Female	170 (52)	194 (57.2)	
Mode of cleaning teeth	Finger	0	10 (2.8)	
	Toothbrush	327 (100)	330 (97.2)	0.084
Frequency of cleaning teeth	Once daily	129 (39.4)	340 (100)	
	Two or more times a day	198 (60.6)	0	<0.001
Material used for cleaning teeth	Toothpaste	327 (100)	320 (94.3)	
	Toothpowder	0	20 (5.7)	<0.05
Type of Diet	Vegetarian	0	46 (13.6)	
	Mixed	327 (100)	294 (86.4)	<0.001
Frequency of sugar consumption	Once a day	-	109 (32.1)	
	Two times a day	196 (59.8)	148 (43.6)	
	Three times a day	131 (40.2)	27 (7.9)	<0.001
	≥ 3 times a day	-	56 (16.4)	
Dental visit	Never visited	327 (100)	234 (68.6)	
	1-3 months back	-	41 (12.1)	<0.001
	4 - 6 months back	-	24 (7.1)	
	> 6 months back	-	41 (12.1)	
Reason for dental visit	Pain with teeth or gums	-	86 (79.5)	
	Routine check up	-	22 (20.5)	<0.001

Table 2: Oral health status of study population

Oral health related behavior variables		Ashrama School N (%)	Govt. School N (%)	p value
Dental Fluorosis	Absent	252 (77.1)	291 (86.4)	<0.001
	Present	75 (22.9)	49 (14.4)	
Periodontal status	Healthy	18 (5.5)	24 (7.1)	0.59
	Bleeding	49 (15)	129 (37.9)	<0.001
	Calculus	260 (79.5)	187 (55)	<0.001
Mean number of sextants	CPI =0	1.07±1.15	1.40±1.54	0.41
	CPI=1	2.92±1.59	3.31±1.58	0.075
	CPI=2	2.00±1.53	1.31±1.53	<0.001
Decayed teeth (DT)	Absent	162 (49.6)	260 (76.4)	<0.001
	Present	165 (50.4)	80 (23.6)	
Missing teeth (MT)	Absent	327 (100)	335 (98.6)	0.16
	Present	-	5 (1.4)	
Filled teeth (FT)	Absent	327 (100)	335 (98.6)	0.16
	Present	-	5 (1.4)	
Mean DMFT		1.15±1.62	0.48±1.04	<0.001
DMFT Range	0-1	162 (49.6)	260 (76.4)	<0.001
	1-3	129 (40.1)	63 (18.6)	<0.001
	≥ 3	36 (10.3)	17 (5)	<0.001
Dental Aesthetic Index scores				
< 25 (No abnormality / minor malocclusion)		260 (80.3)	300 (88.5)	0.45
26 – 30 (Definite malocclusion)		43 (13.4)	24 (7.1)	
30 -35 (Severe malocclusion)		12 (3.1)	11 (2.9)	
36 and Above (Very severe / handicapping malocclusion)		12 (3.1)	5 (1.4)	

Discussion

Society and culture are linked to behavioral patterns or lifestyles. Public health problems are also closely related to the lifestyles of people. The present study explores the association between social disadvantage and oral health.

In the present study, from comparison group none of the children brushed two or more times a day which could be attributed to the attitude of parents and school health policies towards oral health of children. Harikiran et al [11] in a study to assess knowledge, attitude, and practice (KAP) toward oral health among 11 to 12-year-old school children in a government-aided missionary school of Bangalore city reported 38.5% children brushed their teeth two or more times a day.

In Ashrama school 100% children and 67.9% children in comparison group reported of having consumed sugar two or more times the previous day. In National Oral Health Survey and Fluoride Mapping 2002-2003 for 12 year olds 15% had taken sugar two and more times the previous day. High consumption of extrinsic sugar in Ashrama schools was mainly due to the ease of availability in the school environment. Fundamental component of any school based program for promotion of oral health is a healthy environment, with attention to all aspects of the school environment that could affect the health of students.

None of the children from Ashrama schools had visited a dentist. Reason for no dental visits in Ashrama school children was mainly poor socioeconomic conditions of these children, inadequately funded schools, and lack of school based programs for promotion of oral health. Petersen et al [12] in a study done in school children among 12-

year in Southern Thailand reported as high as 66% children saw a dentist within the previous year and 24% reported that visits were due to troubles in teeth.

Higher prevalence of fluorosis in Ashrama school children was due to high fluoride belts in Karnataka State. Calculus was present in 79.5% children from the Ashrama schools as compared to 55% children from the comparison group ($p \leq 0.001$). Per-cent of subjects in 12- year group in National Oral Health Survey 2002-2003 in Karnataka State with calculus was 51.6%, which was similar to the comparison group. Similarily, Penq et al [13] reported a maximum CPITN score of 65% in Chinese urban school children. Despite reporting good oral hygiene practices (higher percentage of children using toothbrush and toothpaste, and claiming brushing two or more times a day), higher percentage of calculus in Ashrama school children could be attributable to improper toothbrushing techniques and lack of individual supervision.

Mean value of Decayed teeth in Ashrama schools and comparison group were 1.15 ± 1.62 and 0.46 ± 0.98 respectively. ($p < 0.001$) The differences could be again attributed to improper oral hygiene practices, dietary factors and no utilization of dental services in Ashrama school children. Jamieson et al [14] described oral health inequalities among indigenous and non indigenous children in the Northern Territory of Australia using an area-based measure of socioeconomic status (SES). Data were obtained from indigenous and non indigenous 4-13-year-old children enrolled in the Northern Territory School Dental Service in 2002-2003. Across all age-groups, socially disadvantaged indigenous children experienced higher mean DMFT levels than their similar age group, similarly disadvantaged non indigenous counterparts, which was also reported by Parker[15].

Prevalence of malocclusion in Ashrama school children and other government schools were similar to findings of National Oral Health Survey 2002-2003 for 12- year olds. Onyeaso et al, [16] in a study in Nigeria reported higher DAI scores in handicapped children than the normal.

In Karnataka, government funded schools charge low fees (Rs.200 – 600/annum), as against private schools charging much higher fees (Rs.5000-10,000/annum). A possible limitation of study could be the socioeconomic "crossover" effect between the two groups.

After the survey, dental treatment camps were organized for Ashrama school children. Survey findings were reported to the Project Coordinator, Indian Tribal Development Project (ITDP), Udupi district.

The study revealed higher levels of dental caries experience, untreated dental disease and social disadvantage of children attending Ashrama schools and provides evidence for the need to address the health inequalities of these children.

The government could appoint a dentist to these schools or at least one dental clinic be set-up for these children, preferably within school premises in all districts. School based programs for promotion of oral health should be initiated with attention being given to all aspects of school environment.

The study finding also indicates the need for closer cooperation between general health and dental professionals. Coordinated efforts from both professionals are required to educate children about the importance of oral health, dietary practices and the relationship of their oral health with their general health. Findings also reflect a need to further explore association between social disadvantage and oral health.

Oral health status among special children

It is estimated that, around 500 and 650 million people worldwide live with a significant impairment. According to the World Health Organization (WHO) 10 per cent of the world's children and young people around 200 million people have sensory, intellectual or mental health impairment. Around 80 per cent of them live in developing countries.[1] Statistics such as these demonstrate that to be born with or acquire impairment is far from unusual or abnormal.

The reported incidence and prevalence of impairment in the population vary significantly from one country to another. Specialists, however, agree on a working approximation giving a minimum benchmark of 2.5 per cent of children aged 0-14 with self-evident moderate to severe levels of sensory, physical and intellectual impairments. An additional 8 per cent can be expected to have learning or behavioural difficulties or both.[2]

The World Bank has estimated that persons with disabilities account for up to one in five of the world's poorest people, that is, those who live on less than one dollar a day and who lack access to basic necessities such as food, clean water, clothing and shelter.[3] These figures have been brought to life in a recent report from Inclusion International which documents the poverty and exclusion experienced daily by people with intellectual disabilities and their families in all regions of the world.[4]

Children with disabilities may be physically, mentally or socially challenged.[5] The term handicap has been defined as a systematic taxonomy of the consequences of injury and disease, according to the International Classification of Impairments, Disabilities and Handicaps (ICIDH).[6] The Maternal and Child Health Bureau (MCHB)[7] has defined children and adolescents with special healthcare needs (SHCN) as those "who have or are at increased risk for a chronic physical, developmental, behavioral, or emotional condition and who require health and related services of a type or amount beyond that required by children generally."

The term Special Needs is a short form of Special Education Needs and is a way to refer to students with disabilities. The term Special Needs in the education setting comes into play whenever a child's education program is officially altered from what would normally be provided to students through an Individual Education Plan which is sometimes referred to as an Individual Program plan. In the USA, special needs is a term used in clinical diagnostic and functional development to describe individuals who require assistance for disabilities that may be medical, mental, or psychological. In the UK, special needs often refers to special needs within an educational context or special educational needs (SEN). [8]

Children with disabilities and their families constantly experience barriers to the enjoyment of their basic human rights and to their inclusion in society. Their abilities are overlooked, their capacities are underestimated and their needs are given low priority. Yet, the barriers they face are more frequently as a result of the environment in which they live than as a result of their impairment.

The additional burden placed on families with children having disabilities, deepens the impact of economic poverty and may further perpetuate discriminatory attitudes towards these groups. Children with special healthcare needs constitute a high-risk group; hence, their health is of importance for the overall development of the society. With half of the world's population under 15 years of age, the number of adolescents and youth with disabilities particularly in developing countries is significantly higher and is on the rise. The National Sample Survey Organization (NSSO) report suggests that the number of disabled persons in the country is estimated to be 18.49 million which forms to about 1.8% of the total population. [9]

Although individuals who are disabled are entitled to the same standards of health and care as the general population, there is evidence that they experience poorer general and oral health, have unmet health needs and lower uptake of screening services. [10] Oral diseases can have a direct impact on the health of children and adolescents with certain systemic health problems or conditions. Children with disabilities may have more marked oral pathologies, either because of their actual disability or for other medical, economic or social reasons, or even because their parents find it very difficult to carry out proper daily oral hygiene (e.g., cariogenic effect of medicines with high sugar content, excessive tooth grinding with self-mutilating behaviors).

Children and adolescents with special healthcare needs (SHCN) have been stated to have higher unmet oral healthcare needs across all income levels. Although numerous studies [11-13] have documented the oral health of children with special healthcare needs, yet there is almost no data available for the WHO advocated index age group of 12 years. Lack of this important data is a serious limitation to oral health comparison of special children and healthy children. This study explores the association of disabilities and oral health. Specifically, the aim of the study was to compare and assess oral health status of twelve- year old special children with same aged healthy children in Karnataka, South India.

Methods

Study design and subjects

The cross-sectional study was conducted among special children of Udupi district Southern India. There are 6 schools for special children in Udupi district. List of the schools along with the strength was obtained from the Women and Child Development Office, Udupi.

A total of 191 children in 12- year age group were enrolled in special schools; all the children were selected and invited to participate in the study. A total of 203, 12 year old were selected randomly for comparison from four other government schools. All the children attending the government schools were in the mainstream of education in the society and were designated as healthy subjects.

The study design consisted of close-ended questions on socio demographic factors, dietary habits, oral hygiene habits, type of disability (in case of children with special healthcare needs) and visits to any health personnel for dental needs was collected by means of personal interviews administered by the examiner. The dental team comprised of the examiner assisted by a recording clerk, an interpreter and a local health worker.

Clinical examination

All the subjects were examined in premises of the respective schools, with tables and portable chairs under adequate illumination (Type III) and clinical data were collected on periodontal status, dental caries, treatment needs and dentofacial anomalies. A pilot study was conducted on 30 children each in comparable age groups to see the feasibility of study and to deduce sample size for comparison group.

Community Periodontal Index (CPI) was used to record the periodontal condition using a mouth mirror and CPI probe. The Community periodontal index (CPI) was introduced by WHO to provide country profiles of periodontal health status and to enable countries to plan intervention programs to reduce prevalence and severity of periodontal disease. For subjects under 20 years, only six index teeth – 16,11,26,36,31 and 46 are examined. This modification is made in order to avoid scoring the deepened sulci associated with eruption as periodontal pockets. For the same reason when children under age of 15 are examined, pockets were not recorded, i.e. only bleeding and calculus scores were recorded.

WHO's criterion was used for detection of dentition status and treatment needs. The examination was conducted with a plane mouth mirror. A systematic approach was adopted for assessment of dentition status and treatment needs. The examination proceeded in an orderly manner from one tooth or tooth space to the adjacent tooth or tooth space. A tooth was considered present in mouth when any part of it was visible. Data on treatment needs are of great value at local and national levels because they provide a basis for estimating personnel requirement and costs of an oral health program under prevailing or anticipated local conditions. Treatment requirements were assessed for the whole tooth, including both coronal and root caries. Dentofacial anomalies were assessed using Dental Aesthetic Index.[14]

Ethical clearance was taken from Kasturba Hospital Ethics Committee, Kasturba Hospital, Manipal. Informed written consent was taken from parents and children before carrying out the survey. The survey was scheduled between the months of Sep 2008 and Jan 2009. All examinations were performed by a single examiner and duplicate examinations were conducted on one of every ten subjects throughout the survey. Intra-examiner reliability for various indices was assessed using kappa statistic which was in range of 0.90 – 0.92.

Statistical analysis

SPSS software version 16 was used for statistical analysis. Mean and standard deviations were calculated for DMFT and their components. Chi Square test was used to compare between categorical variables. Mann – Whitney U-test was used for comparison between two groups for quantitative variables. Logistic and linear regression analysis was performed to determine the importance of the factors associated with caries status. A set of independent variables including type of school attended, gender, frequency of cleaning teeth, frequency of between meal sugar consumption, and utilization of dental care was considered. Odds ratio was calculated for all variables with 95% confidence intervals. All the dependent variables to be included in the regression analysis were dichotomized. $p \leq 0.05$ was considered as statistically significant.

Results

A total of 394 subjects comprised the sample: 191 children with special healthcare needs and 203 healthy controls. The socio-demographic profile of the study population is presented in Table 1. In both study groups, gender was almost equally distributed. Many of the children attending special schools were mentally challenged, with 76 (39.6%) having moderate mental disability (IQ level of 35 to 49). In relation to literacy level, 40% of the mothers of those in the group with SHCN and 38.6% of mothers in the control group had completed middle school and most were not employed (69.1% special and 72.5% control group); 46.6% of the fathers of those in the group with SHCN and 71.9% in the control group had completed high school and 52.8% and 63.6%, respectively, were skilled workers. In both the study groups, the majority of the subjects were above the poverty line (76% and 97% in special and healthy controls respectively). No statistically significant differences were noted between the two groups with respect to demographic variables.

Oral health behavioral characteristics of the study population are presented in Table 2. Statistically significant differences ($p<0.001$) were seen in frequency of sugar consumption between subjects with special healthcare needs and their healthy controls. Brushing frequency in the majority of both groups was once a day, with toothbrush and toothpaste. A total of 160 subjects (83.8%) with special healthcare needs and 179 healthy counterparts (88.1%) reported having never visited a dentist, and there was no statistically significant difference between the groups. In relation to brushing assistance, 46 subjects (24.2%) with special healthcare needs needed help brushing, whereas 201 of the healthy children (99.0%) brushed independently, and there was a statistically significant difference (p< 0.05).

The mean number of sextants with a healthy periodontium, bleeding, and calculus was calculated. Subjects with special healthcare needs had significantly higher CPI scores than their healthy counterparts (p<0.001). One hundred and seventy two (89.8%) 12 year children old special school children had dental caries. In healthy controls caries was present in 119 (58.6%) children. Caries prevalence was found to be higher among subjects with special healthcare needs (p<0.01). The mean values of decayed teeth (DT), missing teeth (MT), and DMFT in subjects with special healthcare needs were found to be higher than for the healthy controls. The D component contributed most to the caries index. Mean DMFT values for special school children and healthy controls were 2.52 ± 2.61 and 0.61 ± 1.12 respectively. There was a significantly higher prevalence of malocclusion in subjects with special healthcare needs, with 66.4% having definite malocclusion, while 82.6% of control subjects had no abnormality/minor malocclusion in healthy controls (p<0.001). The mean values for treatment needs were higher in subjects with special healthcare needs. They had a greater need for fissure sealants, pulp care as well as one- and two-surface restorations (p<0.01). (Table 3)

Table 4 depicts the stepwise multiple linear regression analysis of the caries status (DMFT) in relation to several independent variables, which included school attended, gender, frequency of cleaning teeth, frequency of between meal sugar consumption and dental malocclusion. The variables in the model explained 70% of the variance in caries status for the combined 12- year group. Schools for special children, male gender, low frequency of cleaning teeth, higher in between meal sugar consumption and dental malocclusion were significantly related to dental caries.

Logistic regression analysis was employed to determine the contribution of type of school attended, gender, oral hygiene practices, frequency of between meal sugar consumption, dental visits and dental malocclusion to dental caries. The results of logistic regression showed that all independent variables were significantly related to dental caries. The association between special schools and dental caries was evident with an odds ratio of 2.02 times. Males were more likely to have dental caries, as compared to females with an odds ratio of 0.70. Subjects who cleaned their teeth one or more time a day were less likely to have dental caries then those who cleaned their teeth some times or never (OR = 0.82; P = 0.001). High frequency of between meal sugar consumption was also related to dental caries (OR = 1.01; P = 0.001). Utilization of dental care was inversely related to dental caries (OR = 1.24; P = 0.001). Association was found between malocclusion with dental caries; specifically with severe and handicapping malocclusion (OR = 1.45, P = 0.001). (Table 5)

Table 1: Sociodemographic characteristics of children with special healthcare needs compared to healthy controls

Sociodemographic Variables		Children with special healthcare needs N (%)	Healthy controls N (%)	X^2 (p-value)
Gender	Male	105 (55)	97 (48)	3.2 (0.07)
	Female	112 (45)	166 (52)	
Siblings	0-1	90 (47)	97 (48)	0.98 (0.32)
	≥2	101 (53)	106 (52)	
Maternal Literacy	Illiterate	15 (7.8)	9 (4.2)	2.97 (0.39)
	Completed middle school (7th Grade)	76 (40)	78 (38.6)	
	Completed high school (10th Grade)	67 (35)	104 (51.2)	
	Graduation and higher	33 (17.2)	1 (6)	
Paternal Literacy	Illiterate	8 (4.2)	2 (0.9)	3.97 (0.26)
	Completed middle school	38 (19.8)	40 (19.5)	
	Completed high school	89 (46.6)	146 (71.9)	
	Graduation and higher	56 (29.4)	16 (7.7)	
Maternal Occupation	Nonworking	132 (69.1)	147 (72.5)	0.87 (0.83)
	Non skilled worker	34 (17.7)	52 (25.4)	
	Skilled worker*	21 (11.0)	2 (1.1)	
	Professional	4 (2.2)	0 (0)	
Paternal Occupation	Nonworking	3 (1.5)	1 (0.6)	1.54 (0.67)
	Non skilled worker	77 (40.3)	64 (31.7)	
	Skilled worker*	101 (52.8)	129 (63.6)	
	Professional	10 (5.4)	9 (4.1)	
Family Type	Nuclear **	143 (75)	162 (80)	1.43 (0.23)
	Extended	48 (25)	41 (20.4)	
Family Income	Below Poverty Line ***	45 (24)	7 (3)	3.23 (0.07)
	Above Poverty Line	146 (76)	196 (97)	
Diet	Vegetarian	30 (16)	27 (14)	0.60 (0.43)
	Mixed	161 (84)	176 (86)	

p ≤0.05 - significant ***Skilled:** Special skill, knowledge, or (usually acquired) ability in work.
****Nuclear: a** father, a mother and their siblings in household dwelling.
*****Poverty Line:** expenditure estimated at Rs. 228.9 per capita per month 1993-94 prices ($4.5 US).

Table 2: Distribution of the study population according to oral health behavioral characteristics

Oral health-related behavior variables		Children with special healthcare needs N (%)	Healthy controls N (%)	X^2 (df)	p value
In-between meal sugar consumption on the previous day	Once a day	32 (16)	12 (5.8)	8.9 (2)	<0.001
	Two times a day	159 (84)	147 (72.6)		
	≥ 3 times a day	0	44 (21.6)		
Brushing frequency	Once daily	174 (91.3)	160 (78.8)	6.7 (1)	<0.05
	Two or more times/day	27 (8.3)	43 (21.2)		
Mode of cleaning teeth	Toothpaste	191 (100)	203 (100)	NA	
	Toothpowder	0	0		
Material used for cleaning teeth	Toothbrush	191 (99.6)	201 (99.1)	0.72 (1)	0.40
	Finger	1 (0.4)	2 (1.0)		
Dental Visit	Never visited	160 (83.8)	179 (88.1)	1.21 (2)	0.27
	1–3 months ago	10 (5.3)	13 (6.5)		
	4–6 months ago	14 (7.5)	10 (4.8)		
	>6 months ago	7 (3.4)	1 (0.6)		
Brushing Assistance	Assisted	46 (24.2)	0	12.1 (2)	<0.001
	Non-assisted	102 (53.6)	201 (99.0)		
	Under supervision	43 (22.3)	2 (1.0)		

$p ≤ 0.05$ = significant, NA=not applicable

Table 3: Periodontal disease, Dental caries experience, DAI index and Treatment needs among study subjects

Clinical Variables	CPI score/ Caries experience / DAI scores/ and Treatment needs	Children with special healthcare needs	Healthy controls	p-value
Community Periodontal Index Score (CPI) (Mean sextants ±SD)	0=Healthy	0.1 (0.4)	1.4 (0.9)	<0.001
	1=Bleeding	2.8 (0.9)	2.1 (1.4)	<0.001
	2=Calculus	3.6 (1.5)	2 (0.7)	<0.001
Dental Caries (Absence or Presence) N (%)	DMFT > 0	172 (89.8)	119 (58.6)	<0.001
	DMFT = 0	19 (10.2)	84 (41.4)	
DMFT (Decayed, Missing, Filled) (Mean ±SD)	Decayed (DT)	2.29 (2.49)	0.44 (0.88)	<0.01
	Missing (MT)	0.13 (0.50)	0.01 (0.39)	<0.05
	Filled (FT)	0.10 (0.62)	0.13 (0.49)	0.51
	DMFT	2.52 (2.61)	0.61 (1.12)	<0.01
Dental Aesthetic Index core (DAI) N (%)	< 25 (No abnormality)	43 (22.3)	168 (82.6)	<0.001
	26 – 30 (Definite malocclusion)	126 (66.4)	35 (17.4)	
	30 – 35 (Severe malocclusion)	21 (10.9)	0	
	36 and above (Handicapping malocclusion)	1 (0.4)	0	
Treatment needs (Mean ±SD)	Fissure sealant	0.2 (0.6)	0.1 (0.40)	<0.01
	One surface filling	1.1 (1.3)	0.5 (1.0)	<0.01
	Two or more surface filling	0.9 (1.2)	0.5 (0.8)	<0.05
	Pulp care and restoration	0.8 (1.2)	0.3 (0.7)	<0.01
	Extraction	0.7 (1.3)	0.3 (0.7)	0.32

Table 4: Multiple Linear Regression model for dental caries

Model	R	R^2	Adjusted R^2	SE	R^2 Change	P
1	0.63[a]	0.40	0.40	3.94	.40	0.001
2	0.66[b]	0.44	0.44	3.95	.04	0.05
3.	0.72[c]	0.52	0.52	3.97	.08	0.001
4.	0.77[d]	0.59	0.59	3.94	.07	0.001
5.	0.84[e]	0.70	0.70	3.95	.11	0.001

a. Predictors: School
b. Predictors: School, Gender
c. Predictors: School, Gender, Frequency of Cleaning teeth
d. Predictors: School, Gender, Frequency of Cleaning teeth, Frequency of between meal sugar consumption
e. Predictors: School, Gender, Frequency of Cleaning teeth, Frequency of between meal sugar consumption, Malocclusion

Table 5: Logistic Regression analysis for study population with dental caries as dependent variable (Absence of dental caries, Dt score 0 Vs presence of dental caries, Dt scores ≥1) and School, Gender, Frequency of cleaning teeth, Frequency of between meal sugar consumption and as independent variables

Variables	B	SE B	P	OR (95%CI)
School	0.82	0.0027	0.001	2.02 (1.94, 2.10)
Gender	0.37	0.0021	0.05	0.70 (0.62, 0.78)
Frequency of cleaning teeth	0.62	0.003	0.001	0.82 (0.74, 0.90)
Frequency of between meal sugar consumption	0.65	0.0012	0.001	1.01 (0.93, 1.09)
Dental Visit	0.71	0.0015	0.001	1.24 (1.16, 1.32)
Dental Malocclusion	0.75	0.0026	0.001	1.45 (1.37, 1.53)

Discussion

Designing a system of care for specifically affected children with special health care would require objective data about the actual dental health, such as would be obtained from oral examination. In this study, oral health status and treatment needs of 12 year old children with special healthcare needs were assessed and compared with a group of healthy control subjects who were matched by socio-demographic factors such as age, gender, geographical location, parental literacy, occupation, and family income.

As many as 84% of the children from special schools and 72.6% in the control group reported having consumed sugar two times between meals on the previous day; the difference was highly significant (p<0.001). According to the National Oral Health Survey in India (2002-2003), for 12-year age groups, it was reported that only 24% to 30% of the respondents consumed sugar once the previous day, while 14% to 15% had consumed sugar two or more times. [15]

The majority of the children in both the study groups had never visited a dentist. This may be due to their socioeconomic backgrounds, including family income, parental education, and area residence along with cost of dental care, which might have influenced dental service utilization. De Jongh et al reported that a significantly higher proportion of children with disabilities in their study did not receive any routine dental care in comparison to healthy controls (53.1% and 23.8%, respectively) because noncooperation and communication problems were important barriers leading to a relatively low degree of quality dental care. [16]

Mean CPI scores were significantly higher among children with special healthcare needs compared to healthy controls (p<0.001). This may be attributed to the frequency of brushing, improper tooth brushing techniques, and use of medications in children with special healthcare needs, despite a similar percentage of these children using toothbrush and toothpaste compared to the control group. The findings of this study were similar to those of the National Oral Health Survey among 12-year-olds, where the mean number of sextants with CPI score was 0, 1, and 2 was 1.2, 2.4, and 2.3 respectively. Studies by various authors have also reported significantly greater prevalence of periodontal disease in children with special healthcare needs compared to healthy controls. [11-20]

Caries prevalence was higher in children with special healthcare needs than in the healthy controls, which could be due to poor muscular coordination and muscle weakness interfering with routine daily oral hygiene. Also, frequent use of sugar-sweetened snacks, less frequent brushing, and some socio-demographic factors may be important determinants of caries risk for children in both groups. Similar findings have been reported by other authors [11,21] where the prevalence of dental caries ranged from 78.3% to 89.6% in different types of children with special healthcare needs. Studies by other authors [18,21] have reported higher rates of DMFT than in this study, with values ranging from 4.5 to 12.51 in children with SHCN. Ivancić et al reported the mean DMFT in disabled and healthy children to be 6.39 and 4.76 respectively. [22] Other studies [23-25] reported a mean DMFT of 1.06, 0.8, 2.0, 1 ±1.42 respectively in children with various forms of SHCN, which was lower than our study. A mean DMFT of 1.85 in special school children was seen compared to 1.44 in healthy school children in a study by Shaw et al showing a significantly higher decayed component in children with special healthcare needs than their healthy counterparts, a similar finding as in the present study[19]. The 2002–2003 National Oral Health Survey also reported a lower mean DMFT of 1.87 compared to our study. [15]

There was a significantly higher prevalence of malocclusion in subjects with special healthcare needs as compared to healthy controls. Similarly, other studies have reported normal or minor malocclusions in less than 42%, definite malocclusion in 17% to 24%, severe malocclusion in 9% to 23.6%, and very severe malocclusion in 32% of the children with special healthcare needs. [21,26,27]

Treatment needs for pulp care was higher among children with special healthcare needs, which could be due to untreated caries. The mean values for treatment needs were higher in subjects with special healthcare needs.

Receipt of timely dental services is of particular importance to children with special health care needs because of the higher prevalence of structural irregularities, infections and disease among these children compared with those in the general population. Due to the complications of disability and the cumulative nature of disease, regular training about oral health care is more important for children with special healthcare needs than for healthy children. The dental team should plan on providing comprehensive school-based programs, including oral health education to help children develop skills, provide fluoride supplements and sealants, offer dietary and nutrition counseling, to promote oral health. Professionally made videos featuring both children with special healthcare needs and healthy children with a diversity of contents can be valuable in educating them about oral health.

The provision of oral care for children in the mainstream of society cannot be neglected, as our study also revealed poor use of dental services by these children. The best way to share the responsibility for the healthy development of children of all cultures would be training school staff, family-centered services, community partnerships, and educating public health officials to understand the oral health problems among children. Poor oral health of children with special healthcare needs as compared to their healthy controls in terms of periodontal status, dentition status, treatment needs, and dentofacial anomalies was found in our study, which confirms a need for preventive treatment for these children. Receipt of timely dental services is of particular importance to children with special health care needs because of the higher prevalence of structural irregularities, infections, and disease among these children compared with those in the general population.

Recommendations

From the present research we can say that oral health status observed for Indian school children is poor and there is a high unmet dental treatment needs.

- Children with Internal locus of control appear more likely to engage in positive health and had better oral hygiene and lower dental caries levels. These beliefs may be utilized for planning oral health promotion programs and for formulating advice given to school children. School teachers, parents and health professionals may play a significant role in the development of an internal locus of control from early childhood.
- Knowledge about the periodontal diseases and use of fluoride was found to be low, dental visits were infrequent, and sweet consumption was found to be high among Indian school children. Significant correlation was noted between the Practice-OHIS and Practice-DMFT scores. Apart from providing knowledge it is also therefore essential towards training school children with oral hygiene related practices.
- It is important to sensitize policymakers about spreading health awareness right from school level. Chapters on health education including oral health education must be made mandatory within the school syllabus.
- Among the socially disadvantaged and the special children the study revealed higher levels of dental caries experience and untreated dental disease. The government could appoint a dentist to these schools or at set up at least one dental clinic for these children, preferably within school premises in all districts. School based programs for promotion of oral health should be initiated with attention being given to all aspects of school environment.

- One of the main problems tribal people generally encounter is the introduction of sweets into their daily diets, by multinational companies for example who reach out to such population through sweets, beverages and change their basic diets. This has been confirmed from the present study. Implementation of preventive programs including restriction of sweets in school premises for the school children is the key to good oral health.
- It also indicates the need for closer cooperation between general health and dental professionals. Coordinated efforts by both professionals are required to educate children about the importance of oral health and the relationship of their oral health with their general health. The school authorities should approach nearby dental institutions to give dental services to children. A "**common risk factor approach**" as advocated by WHO may be followed to reduce the risk for both oral diseases and other diseases.

Conclusion

Internal's appear more likely to engage in positive health and had better oral hygiene and lower dental caries levels. These beliefs may be useful for planning oral health promotion programs and for formulating advice given by oral health professionals about their patients' oral health behaviors. Knowledge about the periodontal diseases and use of fluoride was found to be low, dental visits were infrequent, and sweet consumption was found to be high among Indian school children. Among the socially disadvantaged and the special children the study revealed higher levels of dental caries experience and untreated dental disease.

The government has a duty towards its people which needs to be kept in mind when framing the health policies. How we allocate available funds and services will determine the success in meeting the set goals. Separate funds need to be earmarked for oral health in developing countries. The oral health problems of children cannot be neglected and neither can they be blamed for their poor oral health conditions. If we want school aged children to benefit from adequate oral health care, then it needs to be provided through school health programs, where there is easier access.

References

Locus of control as oral health correlates among school children

1. Egan JT, Leonardson G, Best LG, Welty T, Calhoun D, Beals J. Multidimensional health locus of control in American Indians: the strong heart study. Ethnicity & Disease 2009; 19: 338-344.

2. Rotter, J.B. Generalized expectancies of internal versus external control of reinforcements. Psychological Monographs 1966; 80: 609.

3. Chase I, Berkowitz RJ, Pros;,kinHM, Weinstein P, Billings R. Clinical outcomes for early childhood caries (ECC): the influence of health, locus of control. European Journal of Paediatric Dentistry 2004; 5:76-80.

4. Ramos-Gomez FJ, Weintraub JA, Gransky SA, Hoover CI, Featherstone J. Bacterial, behavioural and environmental factors associated with early childhood caries. Journal of Clinical Pediatric Dentistry 2002; 26:165-173

5. Wallston KA. The validity of the multidimensional health locus of control scales. Journal of Health Psychology 2005; 10(5):623–631.

6. Luszczynska A, and Schwarzer R. Multidimensional health locus of control: comments on the construct and its measurement. Journal of Health Psychology 2005; 10(5):633–642.

7. Wallston BS, Wallston KA, Kaplan GD, Maides SA. Development and validation of the health locus of control (HLC) scale. Journal of Consulting and Clinical Psychology 1976; 44(4):580-5.

8. Wallston, K.A., Wallston, B.S., & DeVellis, R. Development of the Multidimensional health Locus of Control (MHLC) Scales. Health Education Monographs 1978; 6:160-170.

9. Cohen M, Azaiza F. Health-promoting behaviours and health locus of control from a multicultural perspective. Ethnicity & Disease 2007; 17(4): 636–642.

10. Brandao IM, Arcieri RM, Sundefeld ML. Early childhood caries: the influence of socio-behavioral variables and health. Locus of control in a group of children from Araraquara Sao Paulo, Brazil. Cadernos de Saude Publica 2006; 22:1247-56.

11. American Dental Association. Based on WHO report series 1962; 1970.

12. Greene. JC, Vermillion JR. The simplified oral hygiene index. Journal of American Dental Association 1964; 68:7-13.

13. World Health Organization. Oral health surveys. Basic methods. Geneva 4th edition 1997.

14. Acharya S. Professionalization and its effect on health locus of control among Indian dental students. Journal of Dental Education 2008; 72 (1): 110-114.

15. Erin L. O'Hea, Bodenlos JS, Moon S, Grothe KB. The multidimensional health locus of control scales: testing the factorial structure in sample of African American medical patients. Ethnicity & Disease 2009; 19(2):192-8.

16. Acharya S, Kalyana PC, Singh S. Influence of Socioeconomic Status on the Relationship between Locus of Control and Oral Health. Oral Health and Preventive Dentistry 2011; 9: 9-16.

17. Lencova F, Pikhart H, Broukal Z, Tsakos G. Relationship between parental Locus of control and caries experience in preschool children –cross-sectional survey. Bio Med Central Public Health. 2008; 12:8:208.

18. Peter K, Bermek G. Oral health: locus of control, health behavior, self-rated oral health and socio-demographic factors in Istanbul adults. Acta Odontologica Scandinavica. 2010 Jan; 69(1):54-64.

19. Wolfe GR, Stewart JE, Maeder LA, Hartz GF. Use of dental coping beliefs and oral hygiene. Community Dentistry and oral Epidemiology.1996; 24:37-41.

20. Skinner, B.F. The behavior of organisms. New York: Appleton-Century-Crofts, 1938.

21. Shehu J and Mokgwathi MM. Health locus of control and internal resilience factors among adolescents in Botswana: a case-control study with implications for physical education. South African Journal for Research in Sport, Physical Education and Recreation. 2008; 30(2): 95-105.

Knowledge Attitude & Practices as oral health correlates

1. Petersen PE, Bourgeois D, Ogawa Hiroshi, Estupinan-Day S, Ndiaye C. The global burden of oral diseases and risks to oral health. Bulletin of World Health Organization 2005; 83: 661-669.

2. Kwan SY, Petersen PE, Pine CM, Borutta A. Health-promoting schools: an opportunity for oral health promotion. Bull World Health Organisation 2005 Sep; 83(9):677-685.

3. Smyth E, Caamaño F, Fernández-Riveiro P. Oral health knowledge, attitudes and practice in 12-year-old schoolchildren. Med Oral Patol Oral Cir Bucal 2007 Dec 1; 12(8): 614-620.

4. Breslow L. From disease prevention to health promotion. JAMA 1999 Mar 17; 281(11):1030-1033.

5. Worthington HV, Hill KB, Mooney J, Hamilton FA, Blinkhorn AS. A cluster randomized controlled trial of a dental health education program for 10-year-old children. J Public Health Dent 2001 Winter; 61(1):22-7.

6. Redmond C, Blinkhorn F, Kay E, Davies R, Worthington H, Blinkhorn A. A cluster randomized controlled trial testing the effectiveness of a school-based dental health education program for adolescents. J Public Health Dent 1999 Winter; 59(1):12-7.

7. Prakash H, Duggal R, Mathur VP, Petersen PE. Manual for multi-centric oral health survey. DGHS, MoHFW, GOI, WHO India; 2004-05.

8. Greene JC, Vermillion JR. The simplified oral hygiene index. J Am Dent Assoc 1964; 68:7-13.

9. Klein.H, Palmer C.E.and Knutson J.W.; Studies on dental caries. Dental status and dental needs of elementary school children. Public health report (Wash) 1938:53:751 -765.

10. Varenne B, Petersen PE, Ouattara S. Oral health behaviour of children and adults in urban and rural areas of Burkina Faso, Africa. Int Dent J 2006 Apr; 56 (2): 61-70.

11. Kassim BA, Noor MA, Chindia ML. Oral health status among Kenyans in a rural arid setting: dental caries experience and knowledge on its causes. East Afr Med J 2006 Feb; 83(2): 100-105.

12. Zhu L, Petersen PE, Wang HY, Bian JY, Zhang BX. Oral health knowledge, attitudes and behaviour of children and adolescents in China. Int Dent J 2003 Oct; 53 (5): 289-298.

13. Peng B, Petersen PE, Fan MW, Tai BJ. Oral health status and oral health behaviour of 12- year old urban schoolchildren in the People's Republic of China. Community Dent Health 1997: 14 (4): 238-244.

14. Petersen PE, Hoerup N, Poomviset N, Prommajan J, Watanapa A. Oral health status and oral health behaviour of urban and rural schoolchildren in Southern Thailand. Int Dent J 2001 Apr; 51 (2): 95-102.

15. Al-Hussaini R, Al-Kandari M, Hamadi T, Al-Mutawa A, Honkala S, Memon A. Dental health knowledge, attitudes and behaviour among students at the Kuwait University Health Sciences Centre. Med Princ Pract 2003 Oct –Dec; 12 (4): 260-265.

16. Christensen LB, Petersen PE, Bhambal A. Oral health and oral health behaviour among 11 -13- year olds in Bhopal, India. Community Dent Health 2003 Sep; 20 (3): 153-158.

Oral health status among socially disadvantaged children

1. Petersen PE. The World Oral Health Report 2003: continuous improvement of oral health in the 21st century – the approach of the WHO Global Oral Health Programme. Community Dental Oral Epidemiol 2003; 31(Suppl.1):3-24.

2. Singh AA, Singh B, Kharbanda OP, Shukla DK, Goswami K, Gupta S. A study of dental caries in school children from rural Haryana J Indian Soc Pedo Prev Dent 1999 Mar; 17(1):24-28.

3. Prasad KVV, Javali SB. Socio demographic and other factors in relation to dental caries experience among adolescents of Karnataka State, India. JIDA 2001; 72:349-51

4. Bali RK, Mathur VB, Talwar PP, Chanana HB. National oral health survey and fluoride mapping 2002-2003 India. New Delhi: Dental Council of India;2004.

7. Retnakumari N. Prevalence of dental caries and risk assessment among primary school children of 6-12 years in the Varkala municipal area of Kerala. J Indian Soc Pedo Prev Dent 1999; 17:135-42.

8. Singh DK. Prevalence of dental caries in school going children of Patna. JIDA 1981; 53:267.

9. Vaish RP. Prevalence of dental caries among tribal school children in Phulbani District, Orissa. JIDA 1983; 55:455-457.

10. Rahmatulla M and Wyne AM. Caries experience and its relation with socioeconomic class in urban school children of India. Indian J Dent Res 1992; 3:1-5.

11. Chen MS. Oral health of disadvantaged populations. In: Cohen LK, Gift HC, editors. Disease prevention and oral health promotion. Socio-dental sciences in action. Copenhagen: Munksgaard; 1995.p. 153-212.

12. Oral health surveys – Basic methods. 4th ed. Geneva: World Health Organization; 1997.

13. Petersen PE, Aarhus NH, Hatyai WA. Oral health status and oral health behaviour of urban and rural school children in Southern Thailand. Int Dent J 2001; 51:95-102.

14. Harikiran AG, Pallavi SK, Hariprakash S; Ashutosh, Nagesh KS. Oral health-related KAP among 11- to 12-year-old school children in a government-aided missionary school of Bangalore city. Indian J Dent Res. 2008 Jul-Sep;19(3):236-42.

15. Peng B, Petersen PE, Fan MW, Tai BJ. Oral health status and oral health behaviour of 12-year old urban schoolchildren in the People's Republic of China. Community Dent Health 1997 Dec; 14(4):238-44.

16. Jamieson LM, Armfield JM, Roberts-Thomson KF. Oral health inequalities among indigenous and nonindigenous children in the Northern Territory of Australia. Community Dent Oral Epidemiol. 2006 Aug; 34(4):267-76.

17. Parker EJ, Jamieson LM. Oral health comparisons between children attending an Aboriginal health service and a Government school dental service in a regional location. Rural Remote Health 2007 Apr-Jun; 7(2):625.

18. Onyeaso CO. Comparison of malocclusions and orthodontic treatment needs of handicapped and normal children in Ibadan using the Dental Aesthetic Index (DAI). Niger Postgrad Med J 2004 Mar; 11(1):40-4.

Oral health status among special children

1. United Nations, Disability Fact Sheet, International Convention on the Rights of Persons with Disabilities, United Nations, New York, 2006.

2. UNICEF, *Children and Disability in Transition in CEE/CIS and Baltic States*, Innocenti Insight, UNICEF Innocenti Research Centre, Florence, 2005.

3. Elwan, A., *Poverty and Disability: A survey of the literature*, Washington, DC: World Bank, 1999. Available from: http://www.un.org/esa/socdev.pdf

4. Inclusion International, *Hear Our Voices: A global report: People with an intellectual disability and their families speak out on poverty and exclusion*, Inclusion International, London, 2006. Available from: www.inclusion-international.org

5. Park K. *Park's textbook of preventive and social medicine*. 19th ed. Banarasi Das Bhanot Publishers, Jabalpur, India: 2007. p. 49,382, 466-469.

6. World Health Organization (WHO). *International Classification of impairments, disabilities and handicaps*. Geneva: World Health Organization; 1980.

7. McPherson M, Arango P, Fox H *et al*. A new definition of children with special health care needs. *Pediatrics* 1998;102:137-140.

8. United Nations Educational, Scientific and Cultural Organization. Strengthening education systems. Available from: http://portal.unesco.org/education/en/ev

9. Disabled Persons in India. National Sample Survey Organisation. Report No. 485 (58/26/1). 2003. p. A1.

10. British Society for Disability and Oral Health. Clinical guidelines and integrated care pathways for the oral health care of people with learning disabilities 2001.

Faculty of Dental Surgery, The Royal College of Surgeons of England, London 2001.

11. Bhavsar JP, Damle SG. Dental caries and oral hygiene amongst 12-14 years old handicapped children of Bombay, India. *J Indian Soc Pedod Prev Dent* 1995;13:1-3.

12. Rao DB, Hegde AM, Munshi AK. Caries prevalence amongst handicapped children of South Canara District, Karnataka. *J Indian Soc Pedod Prev Dent* 2001;19:67-73.

13. Saravanan S, Anuradha KP, Bhaskar DJ. Prevalence of dental caries and treatment needs among school going children of Pondicherry, India. *J Indian Soc Pedod Prev Dent* 2003; 21:1-12.

14. Oral health surveys: Basic methods. 4th ed. Geneva: World Health Organization; 1997.

15. Bail RK, Mathur VB, Talwar PP, Chanana HB. *National oral health survey and fluoride mapping 2002-2003 India.* New Delhi: Dental Council of India; 2004.

16. de Jongh A, van Houtem C, van der Schoof M, Resida G, Broers D. Oral health status, treatment needs, and obstacles to dental care among noninstitutionalized children with severe mental disabilities in The Netherlands. *Spec Care Dentist* 2008;28:111-5.

17. Costello PJ. Dental health status of mentally and physically handicapped children and adults in the Galway Community Care Area of the Western Health Board. *J Ir Dent Assoc* 1990;36:99-101.

18. Ogasawara T, Kasahara H, Hosaka K et al. [Oral findings in severely handicapped patients participating in the periodic dental check-up system for five years--dental caries, gingival recessions and hyperplasias, periodontal diseases.] [Article in Japanese] *Shoni Shikagaku Zasshi* 1990; 28:732-40.

19. Shaw L, Maclaurin ET, Foster TD. Dental study of handicapped children attending special schools in Birmingham, UK. *Community Dent Oral Epidemiol* 1986;14:24-7.

20. Vyas HA, Damle SG. Comparative study of oral health status of mentally subnormal, physically handicapped, juvenile delinquents and normal children of Bombay. *J Indian Soc Pedod Prev Dent* 1991;9:13-6.

21. Shyama M, Al-Mutawa SA, Morris RE, Sugathan T, Honkala E. Dental caries experience of disabled children and young adults in Kuwait. *Community Dent Health* 2001;18:181-6.

22. Ivancić Jokić N, Majstorović M, Bakarcić D, Katalinić A, Szirovicza L. Dental caries in disabled children. *Coll Antropol* 2007;31:321-4.

23. Ohito FA, Opinya GN, Wang'ombe J. Dental caries, gingivitis and dental plaque in handicapped children in Nairobi, Kenya. *East Afr Med J* 1993;70:71-4.

24. Nunn JH, Gordon PH, Carmichael CL. Dental disease and current treatment needs in a group of physically handicapped children. *Community Dent Health* 1993;10:389-96.

25. Kamatchy KR, Joseph J, Krishnan CG. Dental caries prevalence and experience among the group of institutionalized hearing impaired individuals in Pondicherry—a descriptive study. *Indian J Dent Res* 2003;14:29-32.

26. Onyeaso CO. Comparison of malocclusions and orthodontic treatment needs of handicapped and normal children in Ibadan using the Dental Aesthetic Index (DAI). *Niger Postgrad Med J* 2004;11:40-4.
27. Dinesh RB, Arnitha HM, Munshi AK. Malocclusion and orthodontic treatment need of handicapped individuals in South Canara, India. *Int Dent J* 2003;53:13-8.

I want morebooks!

Buy your books fast and straightforward online - at one of world's fastest growing online book stores! Environmentally sound due to Print-on-Demand technologies.

Buy your books online at
www.morebooks.shop

Kaufen Sie Ihre Bücher schnell und unkompliziert online – auf einer der am schnellsten wachsenden Buchhandelsplattformen weltweit! Dank Print-On-Demand umwelt- und ressourcenschonend produziert.

Bücher schneller online kaufen
www.morebooks.shop

info@omniscriptum.com
www.omniscriptum.com

Printed by Books on Demand GmbH, Norderstedt / Germany